Agroforestry Potential of Dacryodes Edulis (G. Don) H.J. Lam

A Tool For Agro Industry Innovation in South West Nigeria

Onyebuchi Patrick Agwu
Adejoke O. Akinyele
Michael Adegboyega Aduradola

London·Istanbul·Moscow·Delhi·Jakarta

Agroforestry Potential of Dacryodes Edulis (G. Don) H. J. Lam:
A Tool for Agro Industry Innovation in South West Nigeria

by Onyebuchi Patrick Agwu, Adejoke O. Akinyele, Michael Adegboyega Aduradola

Printed in the UK

Published by Glimmer Publishing Ltd. Ground Floor 2, Woodberry Grove, London, N12 0DR, England

Glimmer books may be purchased for educational, business, or sales promotional use. Online editions are also available for most titles (*http://glimmerpublishing. com*). For more information, contact our corporate sales department: *sales@glimmerpublishing. com*

April 2018: First Edition

Language: English

See http://glimmerpublishing.com/978-1-78902-004-5 for release details

The Glimmer logo is a registered trademark ofGlimmer Publishing Ltd.
The cover image by Onyebuchi Patrick Agwu
Cover design by Glimmer Publishing Ltd.

ISBN: 978-1-78902-004-5

Dedication

This work is dedicated to my creator who keeps and guides me. I also want to dedicate it to my wife for his love and patience Agwu Oluomachi Elizabeth Patrick and parent Elder and Ezinne P.I. Agwu for their love and prayers

Acknowledgements

I whole heartedly appreciate my Saviour Almighty God , my father and my Lord wholes by the for His grace, mercies kept and protect me and crown me with gifts and good health. I am indeed indebted to my supervisors, Prof. A. M Aduradola and Dr. Adejoke O. AKINYELE for their divine counsel, guidance, important suggestions and reassurance throughout the period of the study. I appreciate my mentor, Prof. B.O Agbeja, who had shown me more love and a keen interest in my career growth and Development. I am also very grateful to the all the lecturer Dr S.O Olajuyigbe, Dr. L.A. Adebisi, Prof. A.O. Oluwadare, Dr. I.O. Azeez, Prof. J.S.A Osho, Prof. A.O Adegeye, , Dr. P.O. Adesoye, , Dr. A.O. Omole and my father's Prof. O.Y. Ogunsanwo and Dr. O.I Ajewole for their positive contribution to my life.

I sincerely appreciate my parents Elder and Ezinne P.I. Agwu, who did their best to give me education. I will never forget my siblings, bro Chido and his family, bro Ugochukwu and his family, bro Uzodinma and his family, Nnamdi and his family, Udo and Ik. I say big thanks to all for your great inspiration to me.

Finally, I appreciate my wife for his love Agwu Oluomachi Elizabeth Patrick who has contributed to my life. May God continue to be with you all.

AGROFORESTRY POTENTIAL OF *Dacryodes edulis* (G. DON) H.J. LAM: A TOOL FOR AGRO INDUSTRY INNOVATION IN SOUTH WEST NIGERIA

Abstract

Dacroydes edulis (G. DON) is a multipurpose tree that provides a dual function of strengthening food security and Carbon sequestration in rural environments. The aim of the study was to investigate the agroforestry potential of D. *edulis* in south west Nigeria and the objective was to access the phytosociology and determine the vegetative propagation strength of D. *edulis* with the view to provide baseline information of its suitability for multi species agroforestry innovation and improve the sustainability of human livelihood in Nigeria.

This study was carried out in the Akinyele and Ibadan North West local government area in Ibadan, an extensive field survey was carried out in the study areas, all compound farms, home gardens and forest were visited in the area.

The data were collected on D. *edulis by* enumerating all the species and also by identifying and enumerating the plant species associated with the subject tree (D. *edulis*). Statistical analysis was done using percentages and charts. For propagation study, two growth hormones, Indole-3-butyric acid (IBA) and Gibberellic acid (GA), were administered at three concentration levels (0ppm, 1000ppm, and 2000ppm), to macotts positioned at three crown levels (upper, middle and lower) on five matured trees. The effects of hormonal treatment and marcotting position on rooting were monitored. The variables assessed include root number, root length, root collar diameter and root biomass. Data collected was analyzed using descriptive and inferential statistics at $p<0.05$ level of significance.

The results show that a total of 66 plant species were found associated with D. *edulis*, the species were of different structures and classes ranging from woody tree, shrub, climbers and herb species belonging to 40 families were encountered in the study area. Many of the species are of high economic and medicial values, cultivated species were fund to be closely associated with D. *edulis*. The study shows that D. *edulis* has the potential coexist with many other species which include the cultivated agricultural crop such as zea mays, Ipomoea, Manihot esculenta and Occimum gratissimum, it can also coexist freely with many woody tree species. The finding from propagation study reveal that there was no significant impacts of hormone type and concentration on the rooting of D. *edulis* marcotts, however, there was significant difference in the marcotting position on the mature tree. The mean root biomass accumulation was 3.85g, 6.83g and 10.37g for the lower, middle and upper levels respectively along the main bole. The mean number of roots for lower, middle and upper were 11.17, 15.48 and 17.49 respectively. The root length of upper layer marcotts was highest (9.56cm) while lower layers produced the lowest (6.15cm). The upper section of the main stem produced the best response to rooting, while the application of hormones may not be necessary for marcotting in D. *edulis*.

Based on the obtained results of this research it may be concluded that D. *edulis* have the great potential of being used in a multi species agroforestry systems, sustainable and can easily be introduced to the farmers that could contributes to regional and local income generation, strengthens food security, improves health care and sequestrate carbon to fight climate change.

Key words: Agroforestry, Phytosociology *Dacryodes edulis,* Hormone, Marcotting, Multipurpose

TABLE OF CONTENTS

Introduction

Agroforestry practices embarked upon by rural farmers and agroforestry experts have been dominated by mono-specific tree components, the population of people who depend on agroforestry output for livelihood is geometrically increasing. This has led to the depletion and disappearance of many valuable trees through deforestation but has also aggravated food insecurity, especially in the rural areas. Multi-species agroforestry improves the sustainability of biodiversity, the production of good quality timber and non timber species, the control of pest and diseases and climate change amelioration. In order to identify the tree species compatible for a multi species agroforestry, the phytosociology needs to be studied. Due to lack of preassessment of the phytosociology of some so called agroforestry trees, lots of agroforestry practices have been neglected. This is because some trees happened to inhibit the growth of other trees and agricultural crops growing around them

Dacryodes edulis is an evergreen humid tropical forest species from Africa and has been introduced in numerous farm settings in Africa particularly where they are planted in association with cash crops like cocoa and coffee, in home-gardens and other agroforestry practices (Ayuk *et al.*, 1999). Its belongs to the family Burseraceae, Common name is African pear and it's also called bush butter tree in some places. It is highly valued for its fruits (Mialoundama *et al.*, 2002) in the Central and West Africa. Vegetative propagation are increasingly becoming more attractive to people because of the added advantage of capturing desirable traits, quicker fruit yields and their potential to resolve problems of seasonality in *D. edulis* fruit production (Mng'omba *et al.*, 2008). Many edible fruits are collected mainly from the wild and their habitats are currently threatened. A proper understanding of the propagations of these species will help the communities and industries on few known arable crops for fruits. Also, knowledge of their propagation methods will help us to understand the appropriate methods to conserves the fruits.
Vegetative propagation is significant because there is too much pressure on the seeds of *D. edulis*, which is the only alternative to achieve inexpensive domestication. In vegetative propagation various factors have been reported to influence the rooting of cuttings in tree crops out of which hormone type and concentration are the most significant (Atangana *et al.* 2006).

There have been few studies to critically evaluate appropriate agroforestry species and the factors of importance in marcotting, the technique produces clonal propagules possessing the same characteristics as those of parent plants and are early fruiting (Kengue *et al.*, 1990; Kengue, 2002). Vegetative propagation should be used to capture and replicate the phenotype of a superior individual mother tree of *D. edulis* with desirable fruit characteristics.

It is necessary to initiate this study aimed at getting information on the phytosociology and vegetative propagation of *D. edulis* and their below and above ground growth attributes. Knowledge on individual tree's below ground growth attributes is very important in ensuring successful establishment of the plantation and their inherent agro-ecosystems services.

Statement of Problem

Macropropagation is very significant as there is too much pressure on the seeds of this species, which is the only alternative to achieve inexpensive domestication. Grafting *D. edulis* trees with scions from adult trees has not produced encouraging results. Some variants of grafting (approach grafting) have produced some success (12%-50%) but need greater skill just like in rooting of cuttings and marcots (Kengue, 2002). In macropropagation various factors have been reported to influence the rooting of cuttings in tree crops out of which hormone type and concentration are the most significant.

Although there have been few studies to critically evaluate the factors of importance in marcotting, this technique produces clonal propagules possessing the same characteristics as those of parent plants and are early fruiting (Kengue *et al.*, 1990; Kengue, 2002). Even though this technique presently gives the best results for *D. edulis* (Kengue, 2002), it cannot be use for mass production since it is done on large branches of *D. edulis* (with diameters of up to 4 to 5 cm) leading to high pruning of trees, and a very low rate of multiplication. However, macropropagation should be used to capture and replicate the phenotype of a superior individual mother tree of *D. edulis* with desirable fruit characteristics.

Justification

The objective of any tree breeder is to evolve a tree with desirable traits giving the most valuable forest products at a very cheap rate and in a short time. This has led to the adoption of fast growing exotic species, which have been widely used for plantation establishment. Nevertheless, the basic fact is that these exotic species can never re-establish the ecological base and dynamics of the tropical rainforest, which the local species have grown with over the years. While it is not wrong to introduce exotic species, there should be judicious use of both indigenous and exotic species thereby complementing each other to satisfy a variety of uses. Hence, there is need for concerted effort in the improvement and conservation of these local species so as to ensure for plantation establishment and *D. edulis* is one of them.

Over the years, there has been serious conflict among land users over land use patterns and this has led to cutting into more and more of the tropical forest leading to serious environmental degradation with its attendant negative consequences. Therefore, an agroforestry package that aims in resolving some of this land use conflict and will alongside provide a wide range of benefits have been recommended. It has been documented by ICRAF that adoption of agroforestry innovations can increase agricultural production on a sustainable basis and hence improve food security for rural people.

The entire plant of *D. edulis* has pharmaceutical properties that are variously exploited by many African communities (Kengue, 2002). *D. edulis* fruit, for example, contains 66% more fat than peanuts (recommended as a high fat food by FAO) (Barany *et al.*, 2004).

The bark of the plant has long been reported in Gabon to treat wounds. In Congo, a concoction of the bark is taken as 43 oral treatments against leprosy and it is also gargled as mouth-wash for the treatment of tonsillitis. In the western parts of Cameroon, the bark is crushed and used in concoctions against dysenteries while in central Cameroon the bark is used to treat toothache (Mapogmetsem, 1994). The leaves made into a plaster have been recently reported to treat snake bites in South West Cameroon (Jiofack *et al.*, 2010). The leaves are also crushed and the resultant juice used to treat skin diseases such as scabies, ring worm, rashes, while twigs from branches are sometimes used as chewing sticks (Igoli *et al.*, 2005; Ajibesin *et al.*, 2008; Okwu and Nnamdi, 2008). *D. edulis* can serve very important purpose hence a more detailed study is needed on its propagation stage so as to establish this fact.

According to agroforestry data base, Germplasm collections have not been initiated for *D. edulis*, nor have any strategies been drawn up for genebanks or in situ reserves for the species.

In view of these benefits therefore, there is need to increase the tree species population through macropropagation in other to ensure their continuous existence for utilization for both rural and urban dwellers.

That notwithstanding, agroforestry providing a dual function of strengthening food security and C sequestration to fight climate change is still little understood. Consequently, it will be necessary to initiate this study aimed at getting insights on the macropropagation propagation methods of *D. edulis* and their below ground growth attributes. Knowledge on individual tree's bellow ground growth attributes is vital in ensuring successful establishment of the plantation and their inherent agro-ecosystems services.

Literature Review

According to Gepts, (2002) domestication can be viewed as the selection process conducted by humans to adapt plants and animals to the needs of humans, whether as farmers or consumers. Interestingly, the process of domestication has been conducted for more than 13,000 years (i.e., since the last ice age), and occurred independently in several regions (Gepts, 2002). Purugganan and Fuller (2009) recognise at least six regions of domestication, including Mesoamercia, the southern Andes (including the eastern piedmonts), the Near East, Africa (probably the Sahel and the Ethiopian highlands), Southeast Asia and China. Sometimes, modern agriculture that clears land for monocultures can exacerbate the process. This probably explains the recent interest in domesticating tree crops from wild forest species in the tropics, as deforestation has increased in proportion to population growth. In Cameroon for example, forest is being destroyed at an annual rate of 1% (FAO, 2007) as a consequence of unsustainable agricultural practices (slash-and-burn shifting cultivation). This accounts for more than 80% of forest cover loss (CARPE, 2005). This has made indigenous and culturally important species a scarce resource, even though they remain much in demand.

Interestingly, some innovative farmers have reacted to deforestation and rarity in the supply of traditional tree products by starting to select and manage useful trees or growing them within their farms. This approach to farmer-driven domestication in which species are brought into a managed environment through planting or retention, is indicative of the conviction that it is worth investing on indigenous fruit species. For example, *Bactrisgasipaes* (peach palm) is an underutilized food crop, cultivated by small-scale farmers in the Amazon forest. Some farmers prefer starchy fruits for flour, while others prefer oily fruits for cooking. Over the years, these farmers have been able to develop visual descriptors to select palms for oil (red waxy coated fruits) and palms for flour (red or yellow not waxy fruits) (Weber *et al.*, 2004). Likewise, anthropic selection of *Vitellaria paradoxa* (sheanut tree) by local farmers in Ghana eliminated unwanted tree species on-farm, leaving only sheanut trees that were selected for tree growth characteristics (tree size, growth, health status, age to fruiting) and yield parameters (nut size, oil content etc.). Similarly, Asaah *et al.* (2003), and Leakey *et al.* (2004) reported that farmers in Cameroon and Nigeria where selecting and multiplying *Irvingia wombolu* and *I. gabonensis* (bush mango) trees respectively, that have 44% large kernels over similar other trees of the same species (particularly in south-eastern Nigeria). Farmers in southern Cameroon have also been reported to select particular trees for their large fruit size as well as other characteristics such as taste and yield (Schreckenberg *et al.*, 2006). Such selective planting by farmers in Cameroon and Nigeria, has been reported by Waruhiu *et al.* (2004), to result in *D. edulis* fruits from trees on farms being 66% larger than those obtain from trees in the wild. These are strategies developed by farmers in order to be self-sufficient for food, micro-nutrients, medicines and their other day-to-day needs (Tchoundjeu *et al.*, 2008). These actions by farmers to retain natural seedlings on farms and in home gardens, and to eliminate trees with products (fruits/nuts) with less desirable attributes as they open up land to cultivate other crops, and the parallel sowing and/or dispersal of seeds of the more delicious fruits they eat, close to the homestead has been suggested to be a form of commensal'domestication. This commensal approach to domestication constitutes one of the building blocks of the pathway to participatory tree domestication. Participatory tree domestication combines agricultural science and technology with traditional knowledge as an integral package (Tchoundjeu *et al.*, 2006). The domestication of agroforestry trees could therefore be considered as a necessary step to promote sustainable agriculture through diversification with species which generate income in local and distant markets, improve diets and health, meet domestic needs, and restore functional agro-ecosystems, as well as empowering local communities.

Domestication Strategies for Agroforestry Trees

Agroforestry practices are widespread in the tropics and used by more than 1.2 billion people (FAO, 2005). These systems generate products that are important for the livelihoods of millions of people in developing countries. The area under agroforestry worldwide has not been exactly determined, but is estimated that over one billion hectares (46%) of farmland have more than 10% tree cover, thus concerning about 30% of all rural people worldwide (Zomer *et al.*, 2009).

In agroforestry systems, different species fulfill diverse functions as providing food, medicine, fodder, timber and income generation from the sales of surplus food stuffs, cash crop products and extracted AFTPs such as fruits, nuts, leaves, bark, etc. to billions of farmers. Even though most of such land use systems are "rich" in species diversity, they are poor in high value economic species, and therefore, hardly contribute to the well-being of the farmers beyond the subsistence level

Much evidence suggests that poverty drives deforestation (Zwane, 2007) and the loss of biodiversity (including primates) (Naughton-Treves *et al.*, 2011). A rise in poverty occurred at the end of the 1980s in most sub-Sahara African countries following recurrent fluctuations in prices of cash crops (cocoa, coffee and rubber) on the world market with disproportionate effects on the more vulnerable rural poor of which women, youths and elderly were particularly hard hit. The World Bank's structural adjustment programme coupled with price liberalization and currency devaluation presented both challenges and opportunities for the rural poor, as these were accompanied, respectively, by price rises for basic commodities and cheapening of potential export products (cash crops). Despite increased export opportunities, the combination of a weak technological environment, and weak price control and regulatory supply mechanisms generally made export crop farmers increasingly uncompetitive and vulnerable. As a result, relatively developed economies in West and Central Africa, like Cameroon, collapsed.

Against this background, there is an urgent need to diversify farmers' livelihood options through the development of sustainable poverty reduction and tree crop management strategies, such as tree domestication. Fruit tree domestication can be considered as a linear process from collection of fruits and seeds in natural forests to cultivation of improved trees species in specialised tree production systems such as monocrop plantations (Wiersum, 2008). This specialisation makes it possible to minimize competition by other crops and optimize commercial production, and thus make the most efficient use of improved tree species. Verheiji (1991) argued that in mixed production systems such as home gardens, its strength lies in its stability rather than peak performance. However, according to Leakey et al., (2005), this perception is changing with increased recognition that the process of tree domestication does not only concern the adjustment (evolution) of tree species characteristics to improved quality and yield to the expectation of users, but also to social and environmental concerns. Thus, while monoculture tree plantations could result in optimal yields when intensively and professionally managed, under other conditions they have several limitations such as increased susceptibility to pests and diseases and/or, limited biodiversity among others.

Vegetative Propagation

Vegetative propagation can be considered as a phenomenon of regeneration of differentiated, somatic cells of plants to produce a new organ and restore body parts that have been lost by injury or autotomy (selfamputation of body parts) (De Klerk, 2000). Vegetative propagation is defined as the regeneration of new individuals from vegetative organs such as stems, roots, leaves, buds and even single cells. Plants have a remarkable capability to replace lost parts and may even grow a complete, fully new organism from a single somatic cell cultured *in vitro*. De Klerk (2000), distinguishes three types of regeneration in plants: (1) caulogenesis (adventitious shoot formation); (2) rhizogenesis (adventitious root formation); and (3) somatic embryogenesis (adventitious embryo formation).

For the purpose of producing plants that are true-to-type, the domestication strategy adopted in agroforestry tree domestication is the clonal propagation approach based on well-known horticultural techniques of vegetative propagation (Leakey, 2004; Tchoundjeu *et al.*, 2000) applied in a simple, robust and low-tech manner (Leakey *et al.*, 1990), so as to be appropriate for implementation in remote areas of tropical countries which lack reliable supplies of running water or electricity but also other basic resources.

Vegetative propagation is a powerful means of capturing existing genetic traits and fixing them over generations so that they can be used as the basis for a genetic variety'or cultivar'development process. The advantage of using clonal propagules outweighs those of seedlings especially when the products are of high nutritional or income value, or when the tree has a long juvenile phase before first fruiting, and when seeds are scarce, difficult to germinate or difficult to store (Leakey and Akinnifesi, 2008; Tchoundjeu *et al.*, 2006). The resultant uniformity in the eventual crop is advantageous in terms of maximizing quality, matching market specifications and increasing (if this has been a selection criteria) productivity, but it also increases the risks of pest and disease problems. Therefore, risk aversion through diversification of initial clonal production population is a crucial component in the domestication strategy adopted. However, through agroforestry, risk aversion can also be achieved by diversification of the agro-ecosystem by introducing other species and food crops in order to improve the overall agro-ecological system functions (Leakey, 2010)

Vegetative propagation is the reproduction from vegetative parts of the plant. Plant organs could be excised and rooted in a suitable medium for root and shoot formation. This is possible because every cell of the plant contains the genetic information necessary to regenerate the entire plant (Teklehaimanot *et al.*, 2007)

Many new plants can be started in a limited space from a few stock plants. It is possible for reproduction to occur through the formation of adventitious roots and shoots or through the uniting of vegetative parts (Jaenicke and Beniest, 2002). It is useful in the production of cultivars that are seedless, promising species which have insufficient supply of seeds due to mammalian predation, pest and disease attack.

Vegetative propagation of plant species may take advantage over time after a slower initial growth. It may outgrow the seedlings. Vegetative propagation involves isolation of certain plant parts such as stem, root or leaves and introducing them to develop into separate individuals. It is a very useful technique for maintaining and preserving genetic characteristics (Hendromono, 2007). In propagation by cuttings, a portion of stem, roots or leaf is cut from parent plant, after which it is placed under favourable environmental conditions and induced to form roots and shoots, thus producing a new independent plant which in most cases, is identical to the parent plants.

Vegetative propagation especially with stem cuttings has been very successful and popular with many indigenous forest hardwoods. The use of stem cuttings could be by single or multiple nodes and the presence of leaf is very important (Aminah *et al.*, 2006). Stem cutting is simple, rapid, requires less space and relatively inexpensive. It does not require special skills necessary for grafting or budding. It is the most acceptable method for raising clonal planting stock (Hartman *et al.*, 2014).

In cutting, there is no problem of compatibility or poor graft unions. Greater uniformity is obtained by absence of the variation, which sometimes appears as a result of the variable seedlings rootstocks of grafted plant. The parent plant is usually reproduced exactly with no genetic change. Though most trees can be propagated by grafting, this is too expensive for use in forestry. Through research and appropriate technology, feasible methods for mass propagation by cuttings of some important forest trees have been developed. Such species include *Triplochiton scleroxylon, Terminalia ivorensis and Pinusspp.*

Hartman *et al.* (2014), after comparing the growth rate of clonal stocks from stem cuttings and seedlings reported that seedlings only survived best over the initial nine months.

Propagation by Stem Cuttings

Cuttings are portions of stems, roots or leaves that are detached from plants and used to clonally multiply new plants (Hartmann *et al.*, 2002). Two major groups of cuttings can be distinguished: i.e. soft and hardwood cuttings. As a result of morphological differences among these two types of cuttings, the factors responsible for their rooting differ. For example, leafy stem cuttings of softwood species depend on photosynthates produced during propagation conditions, whereas leafless hardwood cuttings rely on hydrolysis of carbohydrates stored within the stem (Leakey,

2004). In a review of the physiology of vegetative propagation, Leakey (2004) identified propagation environment, post severance treatment, stock plant factor and management as key determinants for the rooting ability of cuttings under propagation.

Propagation Environment

These are essentially the enabling conditions within a propagation system that encourage physiological activities (photosynthesis and transpiration) in leaves by minimizing stress of leaf tissues from transpiration and respiration, while also ensuring meristematic activities (mitosis and cell differentiation) prevail in the stem.

Meristematic differentiation is a prerequisite to adventitious root formation and to successful rooting of cutting under propagation. Commercial propagators successfully regulate environmental conditions to maximize rooting (i.e. intermittent mist and/or fog systems, temperature and light manipulation) (Hartmann *et al.*, 2002). A low tech propagation system, non-mist propagation, developed in the 80s is currently widely used in rural communities in West and Central Africa for successful cutting multiplication. Non-mist propagators are constructed following a design based on that of Howland (1975), further modified as described by Leakey and Longman (1987) and Leakey *et al.* (1990). The basic principle underlying all these propagation systems is need to make water available at both the base and leaf of the cutting in a shady cool environment with low vapour pressure deficit (VPD) as this will minimize stress from water (Leakey, 2004).

Propagation by Marcotting (Air Layering)

Marcotting (or air layering), is a technique in which an aerial stem is girdled and enclosed in a rooting medium to produce roots on the upper part of branch while still on the tree (Hartmann *et al.*, 2002). This technique has been developed and used successfully for the propagation of *D. edulis* (Mampouya *et al.*, 1994; Kengue and Tchio, 1994). Mialoundama *et al.* (2002), recommended that *D. edulis* marcots should be set on horizontal branches with diameter above 4 cm. cited by Kengue (2002), reported on the contrary that irregular rooting was observed to characterize marcotted plants derived from horizontal and sub-horizontal branches of *D. edulis*.

In a recent study on the effect of branch orientation (orthotropic, oblique or plagiotropic branches) on root distribution around the base of rooted marcots of *D. edulis*, it was deserved that plagiotropic and oblique branches (<450 from the orthotropic branch) have a greater tendency to develop irregularly distributed roots (ICRAF, unpublished data).

Although there have been few studies to critically evaluate the factors of importance in marcotting, this technique produces clonal propagules possessing the same characteristics as those of parent plants and are early fruiting (Kengue *et al.*, 1990; Kengue, 2002). Even though this technique presently gives the best results for *D. edulis* (Kengue, 2002), it cannot be used for mass production since it is done on large branches of *D. edulis* (with diameters of up to 4 to 5 cm) leading to high pruning of trees, and a very low rate of multiplication. However, it should be used to capture and replicate the phenotype of a superior individual mother tree of *D. edulis* with desirable fruit characteristics

.

Ecology and Geographic Distribution

D. edulis is a shade-loving species of non-flooded forests in the humid tropical zone. It can develop under light shade (hemi-sciaphile) but prefers open areas (heliophile) (Kengue, 2002). According to the latter author, temperature and rainfall are the two major climatic factors that influence growth and development of the tree. The tree develops well under an average temperature range of 23 - 25 °C, and with an annual rainfall range of 1,400 - 4,000 mm. Very high rainfall encourages vegetative development to the detriment of fruit production. Average altitude for best performance is 1,000 m.

D. edulis grows on various soil types. Nonetheless it prefers slightly acidic, deep ferallitic and evolved volcanic soils with exploitable thick and humic horizons (Kengue, 2002). *D. edulis* can be cultivated widely, since it adapts well to differences in day length, temperature, ainfall, soils and altitude. It is planted in southern Nigeria, Cameroon and Democratic Republic of Congo for its nutritious fruit, which has high oil content.

D. edulis is a native to Angola, Benin, Cameroon, Central African Republic, Congo, Cote d'Ivoire, Democratic Republic of Congo, Equatorial Guinea, Gabon, Ghana, Liberia, Nigeria, Sierra Leone, Togo, Uganda and exotic to Malaysia.

Materials and Methods

Study One: Phytosociology study of *D. edulis*

This study was carried out in the Akinyele and Ibadan North West local government area in Ibadan, an extensive field survey was carried out in the study area, all compound farms, home gardens and forest were visited in the area. These forests are located around the school's administrative and residential area. Akinyele is a Local Government Area in Oyo State, Nigeria. It is one of the eleven local governments that make up Ibadan metropolis (latitude 7°28'N, longitude 30°52'N, altitude 277m). The climate is the West Africa monsoon with dry and wet seasons. The dry season lasts usually from November through March and is characterized by dry cold wind of harmattan. The wet season usually lasts from April to October with occasional strong winds and thunderstorms. The annual rainfall in the area is 1258 mm - 1437 mm with mean daily temperature ranging from 22°C - 31oC. Soil type is ferric luvisols. The collection of data was accomplished through identification, enumeration and measurement of distances between the subject tree (*D. edulis*) and its associates' species. The closer a species is to *D. edulis*, the more associated is the species with the subject tree. To identify the species associated with the subject tree, a search radius method according to Sabiiti and Cobbina (Sabitti En, *et al* 1992) was used. Crown diameter of the subject tree was measured using 50m tape for the estimation of the search radius. An associate species is a single or multi-stemmed individual located within the search radius of the subject tree (Sabitti En, *et al* 1992). Search radius (SR) is the distance from the subject tree within which all other species are considered associate species. It can be calculated as: SR = 7/4 × CD, where CD is a crown diameter of the subject tree. Data were analyzed using Frequency, percentages and graphs. Past statistical software used for the data analysis.

Study Two: Macropropagation of *D. edulis*

The study was carried out at Central Nursery Forestry Research Institute of Nigeria (FRIN) FRIN is located on the longitude 07023'18"N to 07023'43"N and latitude 03051'20"E to

03051'43"E. The climate of the study area is the West African monsoon with dry and wet seasons. The dry season is usually from November through March and is characterized by dry cold wind of harmattan. The wet season usually starts from April to October with occasional strong winds and thunderstorms. Mean annual rainfall is about 1548.9 mm, falling within approximately 90 days. The mean maximum temperature is 31.90C, minimum 24.20C while the mean daily relative humidity is about 71.9% (FRIN 2015).

Hypotheses

(1) Null hypothesis (H_0): there are no significant differences between the two hormonal type, three concentrations level and three crown levels of *D. edulis* macortts and cuttings

Methodology

To determine the response of *D. edulis* macortts to hormone type, hormonal concentrations and marcotting position

Air Layering (Marcotting): Equal diameter horizontal branches with thick bark were used for air layering (or marcotting) of *Dacryodes edulis*. Growth hormones (IBA, GA) were applied at three concentrations each (0ppm, 1000ppm and 2000ppm). The application of growth regulators was expected to accelerate rooting and reduces the severance period of rooting, the marcotting were set at three positions on the crown (lower, middle and upper levels), the rooting medium used was sawdust.

The three factors are hormone types, hormonal concentrations and marcotting positions (lower, middle and upper levels)

Data Collection

The variables collected for the experiments include the roots count, roots length, roots diameter and roots biomass.

The Experiment Design and Data Analysis

A 2*3 factorial experiment in completely randomized design (CRD) were used to set up the experiment.

Factor 'H' is Hormone type in two levels (Indole-3-butyric acid (IBA), Gebreliline (GA)

Factor 'M' Marcotting Position which is in three levels (upper, middle and lower)

The statistical model is $Y_{ijk} = \mu + H_i + G_j + C_k + HM_{ij} + HM_{ik} + MC_{jk} + HMC_{ijk} + E_{ijk}$

Where,

Y_{ijk}= observation made in the ith Hormonal type, j^{th} Marcotting level and k^{th} Hormonal concentration.

μ = General mean effect common to all effects.

H_i = main effect of Hormonal type.

G_j= main effect of Marcotting position.

C_k = main effects of hormones concentration.

HG_{ij} = Interaction between hormone type and Marcotting position.

HC_{ik} = Interaction between hormone type and hormones concentration.

GC_{jk} = Interaction between Marcotting position and hormones concentration.

HGC_{ijk}= Interaction between hormone type, Marcotting position and hormones concentration.

E_{ijk} = error term association with the ijk[th] experiment.

Results and Discussion

Phytosociology Potential of *D. edulis*

A total of 66 plant species were found associated with *D. edulis*, the species were of different structures and classes ranging from woody tree, shrub, climbers and herb species belonging to 40 families (Table 1) were encountered in the study area. Many of the species are of high economic and medicial values, cultivated species were fund to be closely associated with *D. edulis*. The study shows that *D. edulis* has the potential coexist with many other species which include the cultivated agricultural crop such as zea mays, Ipomoea, Manihot esculenta and Occimum gratissimum, its also coexist freely with many woody tree species. For the herbs and shrub, Tridax procumbens had the first associate with the frequency of 52 and occupy about 5.1% of the area occupied by associate species, Axonopus compressus, Peperomia pellucid, Ageratum conyzoides, Desmodium adscendens, Chromolaena odorata and Solenostemon monostachyus were also highly associated with frequencies of 45,39,36,34,31 and 29 respectively and they occupy about 4.4%, 3.6%, 3.4%, 3.1% and 2.9% respectively. The herbs and shrubs species that have weakest associated with *D. edulis* are Sida acuta, Phyllanthus nuriri, Fluerya aestuans, Occimum gratissimum and Combretum. Spp had frequency and percentage cover of 13 (1.3%), 12 (1.2%), 11(1.1%) 8 (0.8%) and 3 (0.3%). For woody tree species, Albizia zygia and Alchornea cordifolia had the highest associate with the frequency of 33 each and they also occupy about 3.3% of the area occupied by associate species, Morinda lucida, Newbouldia laevis, Citrus sinensis and Bridelia micrantha also associated with *D. edulis* with the proportion of 2.2% , 1.7% and 1.5% respectively. On the other hand, wood tree species that have the weakest association with *D. edulis* are Milicia excels, Sterculia tragacantha, Anthocleista djalonensis, Persea Americana, Gneliana arborea and Terminalia Mentalis with the proportion of 0.9%, 0.7%, 0.6%, 0.3% and 0.2% respectively. Based on the frequency and percentage of occurrence of the species, Tridax procumbens had the highest population of individual species 5.2% of the total species population. This is followed by Axonopus compressus and Ageratum conyzoides, while Gneliana arborea and Terminalia Mentalis had the lowest population of individual species with 0.3% and 0.2% respectively.

The percentage distribution of species into families (Figure 2) shows that Asteraceae (15.5%) is the highest represented family. This is closely followed by Euphorbiaceae (10.6%). Next to Euphorbiaceae are Rubiaceae, Poaceae and Moraceae with (6.4%, 5.9% and 4.6%) representation and the lowest family represented are Bombaceae and Cealsapineaceae with 0.3%

Species	Local / Common Name	Family	Frequency	Proportion (%)
Abelmoschus esculentus (L.) Moench	Okra	Malvaceae	12	1.2
Adenia cissampeloides (Planch. ex Hook.) Harms	Adenia	Passifloraceae	10	1.0
Ageratum conyzoides var hirtum (Lam.) DC.	Goat week	Asteraceae	36	3.6
Albizia saman (Jacq.)		Mimosaceae	10	1.0
Albizia zygia J. F. Macbr	Banabana	Mimosaceae	33	3.3
Alchornea cordifolia (Schum. & Thonn.) Müll.	Christmass bush	Euphorbiaceae	25	2.5
Alstonia boonei De Wild	Stool wood	Apocyneceae	11	1.1
Annona muricata L.	Sour sop	Annonaceae	16	1.6
Anthocleista djalonensis A.Chev	Cabbage tree	Loganiaceae	6	0.6
Antiaris africana Engl.	False iroko	Moraceae	13	1.3
Asystasia gangantica (L.) T.Anderson	Senegal tea tree	Acantheceae	25	2.5
Axonopus compressus (Sw.) P.Beauv.	Tropical carpet grass	Poaceae	45	4.4
Azadirachta indica A. Juss.	Neem	Miliaceae	10	1.0
Blighia sapida (Lovett).	Akeeaple	Sapindaceae	14	1.4
Boerhaavia diffusa Linn	Hogweed	Nyctaginaceae	24	2.4
Bombax buonopozense P.Beauv.	Red cotton tree	Bombaceae	4	0.4
Borreria ocymoides	Irawo ile	Rubiaceae	19	1.9
Bridelia micrantha (Hochst.) Baill	Mitzeeri sweetberry	Euphorbiaceae	15	1.5
Carica papaya L.	Pawpaw	Caricaceae	8	0.8
Chromolaena odorata (L.) R.M.King & H.Rob.	Siam weed	Asteraceae	31	3.1
Chrysophyllum albidum G.Don	African star apple	Sapotaceae	12	1.2
Citrus aurantifolia Christre. & Panzer	Swing	Rutaceae	9	0.9
Citrus sinensis L.	Sweet orange	Rutaceae	17	1.7
Cocos nucifera L.	Agbon/ coconut	Palmae	4	0.4

	palm			
Combretum. spp		Combretaceae	3	0.3
Cyathula prostrata L.	Pasture weed	Amarantheceae	17	1.7
Delonix regia (Boj. ex Hook.) Raf.	Flame of th forest	Leguminosaceae	13	1.3
Desmodium adscendens		Leguminosaceae	34	3.4
Eleais guineensis Jacq.	Oil Palm	Palmae	8	0.8
Ficus capensis hunb	Fig tree	Moraceae	13	1.3
Ficus exasperata Vahl	Sand paper tree	Moraceae	12	1.2
Fluerya aestuans (L.) Chew	African nettle	Urticaceae	11	1.1
Gneliana arborea Roxb. ex Sm	Gniliana	Verbanaceae	3	0.3
Ipomoea involucrata	Morning glory	Convolvulaceae	19	1.9
Irvingia gabonensis (Aubry-Lecomte ex O'Rorke) Baill.	Wild mango	Irvingiaceae	9	0.9
Manihot esculenta Crantz	cassava	Euphorbiaceae	15	1.5
Mangifera indica L.	Mango tree	Anacardaceae	4	0.4
Milicia excelsa (Welw.) C.C. Berg	Iroko	Moraceae	9	0.9
Morinda lucida Benth.	Oruwo	Rubiaceae	22	2.2
Moringa oleifera Lam.	Moringa tree	Moringaceae	10	1.0
Newbouldia laevis (ogilisi)	Dants	Bignoniaceae	20	2.0
Occimum gratissimum (L.)	Balsam	Labiatae	8	0.8
Oldelandia corymbosa Linnaeus.	Oyigi	Rubiaceae	24	2.4
Peperomia pellucida Kunth	Silver bush	Piperaceae	39	3.8
Persea americana Mill	Avocado pear	Lauranceae	5	0.5
Phyllanthus amarus Schumach. & Thonn.	Eyin olobe	Euphorbiaceae	23	2.3
Phyllanthus nuriri L.		Euphorbiaceae	12	1.2
Pinus cariabia Morelet	Pine	Pinaceae	12	1.2
Polyalthia longifolia Sonn.	**Police Tress**	Annonaceae	5	0.5
Psidium guajava L.	Guava	Myrtaceae	12	1.2
Rauvolfia vomitoria Afzel.	Swizzler stick	Apocyneceae	13	1.3
Securinega virosa omm. ex A.Juss.	Iranje	Euphorbiaceae	17	1.7
Senna siamea (Lam.) Irwin et Barneby		Cealsapineaceae	4	0.4
Sida acuta Burm.f	Hornbeans leaf	Malvaceae	13	1.3
Solenostemon monostachyus (P Beauv.) Briq.	Catnip	Libiatae	29	2.9
Spondias mombin L.	Hog plum	Anacardaceae	8	0.8
Sterculia tragacantha Lindl.	Star chesnut	Steculiaceae	7	0.7
Synedrella nodiflora (L.) Gaertn.	Tanaposo	Asteraceae	21	2.1
Talinium tangulea	Water leaf	Talinaceae	14	1.4
Tectona grandis L.f.	Teak	Verbanaceae	11	1.1
Terminalia catappa L.	Almond tree	Combretaceae	13	1.3
Terminalia Mentalis L.	Terminalia	Combretaceae	7	0.7
Thevetia neriifolia Juss.	Bush milk	Apocyneceae	2	0.2
Tridax procumbens L.	Tridax	Asteraceae	52	5.1

Vernonia amygdalina Delile	Bitter leaf	Asteraceae	17	1.7
zea mays L.	Maize	Poaceae	15	1.5

TABLE 1. Tree species associated with *D. edulis*

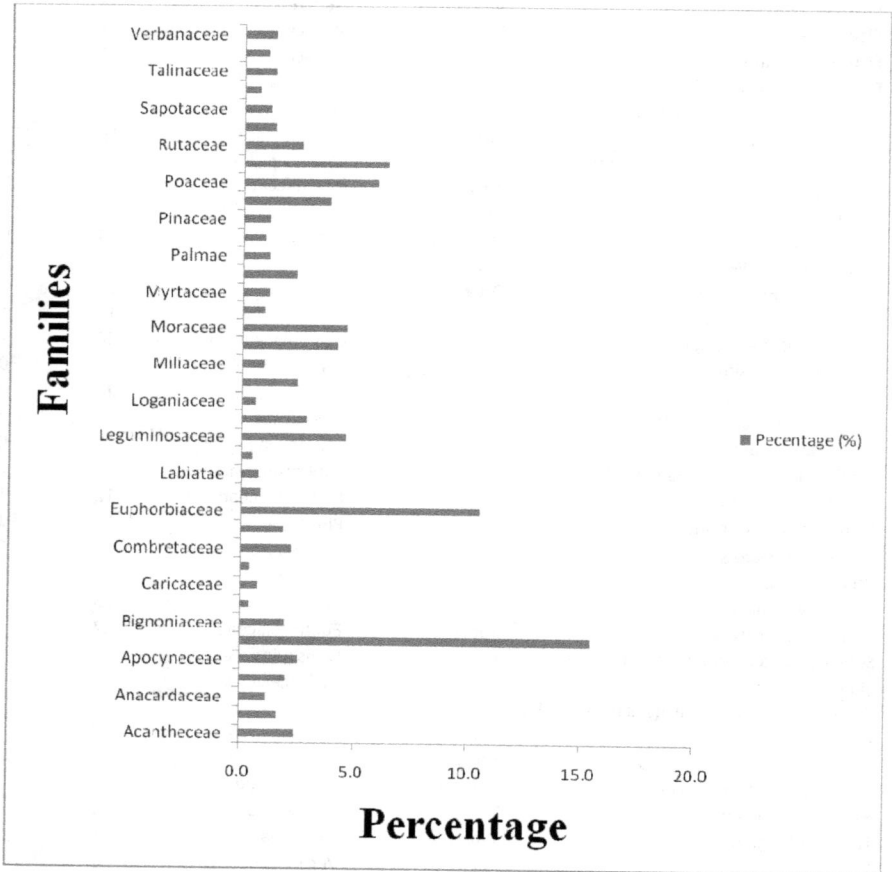

Figure 1. The distribution of the *D. edulis associates* by family

	Number of macortts	GA 1000PPM (Germination %)	GA 2000PPM (Germination %)	GA CONTROL (Germination %)	IBA 1000PPM (Germination %)	IBA 2000PPM (Germination %)	IBA CONTROL (Germination %)
Upper	90	94	93	93	92	94	93
Middle	90	82	84	82	83	83	81
Lower	90	76	78	74	77	77	74
Mean	90	84	85	84	84	85	83
Total	370	252	255	249	252	254	248

TABLE 2: The Rooting percentage of *D. edulis* macortts to hormone type, hormonal concentrations and marcotting position

The shows that *D. edulis* have a very good rooting ability when propagated by mean vegetative propagation (macortting). The vegetative propagation of *D. edulis* looks very promises, as the rooting percentage is very high in the positions where the macortts were set.

The response of *D. edulis* macortts to hormone type, hormonal concentrations and marcotting position

Macortts were allowed to grow for three months. The variables were assessed for another two months, the variables include roots number, roots colar diameter, roots length and roots biomass. (Table 1). After two months of assessment, roots number at the upper section of the crown gave the highest mean values (Table 1), marcotts tested with IBA at 1000ppm had 12.53, IBA at 2000, 12.37, GA control had 12.36, GA at 1000ppm had 12.12, GA at 2000ppm had 12. 27, the lower section of the crown gave the lowest mean values despite the same hormonal treatment on them, IBA at 2000ppm had 8.55, IBA at 1000ppm had 8.33, IBA control had 8.53, GA at 2000ppm had

8.53, GA at 1000ppm had 8.40 and GA control 8.35. Result shows that, for roots length upper section of the crown gave the highest mean values (table 1) IBA at 2000ppm had 7.71cm, IBA at 1000ppm had 7.72cm, IBA control 7.72cm, GA at 2000ppm had 7.72cm,

GA at 1000ppm had 7.72cm and GA control had 7.69cm. The lower section of the crown also gave the lowest mean values: IBA at 2000ppm had 5.15cm, IBA at 1000ppm had 5.28cm, IBA control had 5.13cm, GA at 2000ppm had 5.12cm, GA at 1000ppm had 5.11cm and GA control had 5.12cm. Assessment of roots Biomass shows that upper section of the crown gave the highest mean values: IBA at 2000ppm had 4.76cm, IBA at 1000ppm had 4.65cm, IBA control had 4.81cm, GA at 2000ppm had 4.72cm, GA at 1000ppm had 4.71cm and GA control had 4.65cm. The lower section of the crown gave the lowest mean values: IBA 2000ppm 2.47cm, IBA 1000ppm 2.46cm, IBA control 2.41cm, GA 2000ppm 2.41cm, GA 1000ppm 2.43cm and GA control 2.45cm.

Variables	Marcotting Position	GA 1000PPM	GA 2000PPM	GA CONTROL	IBA 1000PPM	IBA 2000PPM	IBA CONTROL
Root No	Lower	8.40	8.53	8.35	8.33	8.55	8.53
	Middle	11.38	11.23	11.18	11.30	11.59	11.32
	Upper	12.12	12.27	12.36	12.53	12.37	12.31
Roots Lenght (cm)	Lower	5.11	5.12	5.12	5.28	5.15	5.13
	Middle	6.00	5.99	5.98	5.99	5.98	5.96
	Upper	7.72	7.72	7.69	7.72	7.71	7.72
Roots Colar diameter (cm)	Lower	2.94	2.97	2.95	2.94	2.96	2.94
	Middle	3.23	3.24	3.24	3.28	3.26	3.23
	Upper	3.61	3.36	3.64	3.64	3.69	3.58
Roots Biomass (cm)	Lower	2.43	2.41	2.45	2.46	2.47	2.41
	Middle	3.45	3.46	3.43	3.51	3.51	3.41
	Upper	4.71	4.72	4.65	4.65	4.76	4.81

TABLE 3: Mean roots number, roots collar diameter, roots length, roots biomass, hormone types (GA and IBA), concentration and marcortting position

Marcotting position	Mean root length	Mean root collar diameter	Mean root number	Mean root biomass
Lower	5.10[a]	2.92[a]	8.35[a]	2.43[a]
Middle	5.92[b]	3.24[b]	11.36[b]	3.61[b]
Upper	7.70[c]	3.59[c]	12.28[c]	4.72[c]
p-value	0.000*	0.000*	0.000*	0.000*

Means with the same alphabet are not significantly different from each other $p < 0.05$
*=significant ($p < 0.05$)

Table 4: Duncan multiple test to compare the different between pair of means on the effects of Marcotting position on mean root length, collar diameter, root number and root biomass of *D. edulis*

The result from (Table 4) shows that there is significant difference between the marcotting Position on the root length, root collar diameter, root number and root biomass of *D. edulis*. The upper section has the highest mean value for all the variables assessed, in root length upper section had 7.70cm and lower section has the lowest of 5.10cm.

In root collar diameter of *D. edulis* the upper section has the highest mean value of 3.59cm and lower section has the lowest of 2.92cm. Root number of D. *edulis* the upper section has the highest mean value of 12.28cm and lower section has the lowest of 8.35cm, and in root biomass of *D. edulis* the upper section has the highest mean value of 4.72cm and lower section has the lowest of 2.43cm.

Roots length

The data obtained was subjected to analysis of variance (ANOVA)

The result from statistical analysis shows that there were no significant differences ($p > 0.05$) in the hormone types (GA and IBA) and at differences concetntrations., but there were significant differences ($p > 0.05$) in the marcortting position (upper, middle and lower) (Appendix 1)

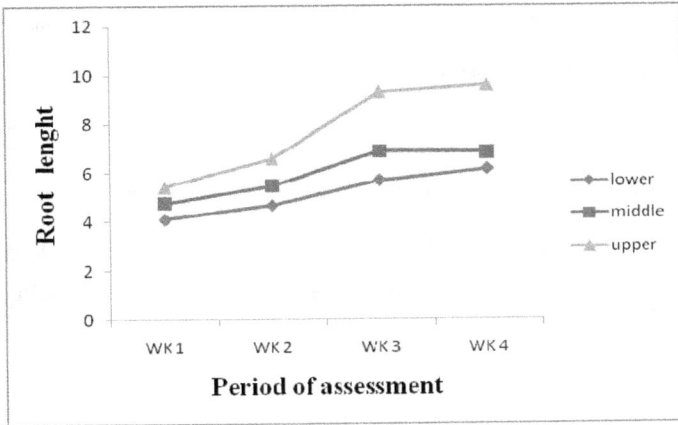

Figue 2 : the response of _D. edulis_ root length macortts to marcotting position

From (figure 2) the graph shows the weekly mean increament of root length, its shows a clear distinction between the marcotting position (lower, middle and upper).

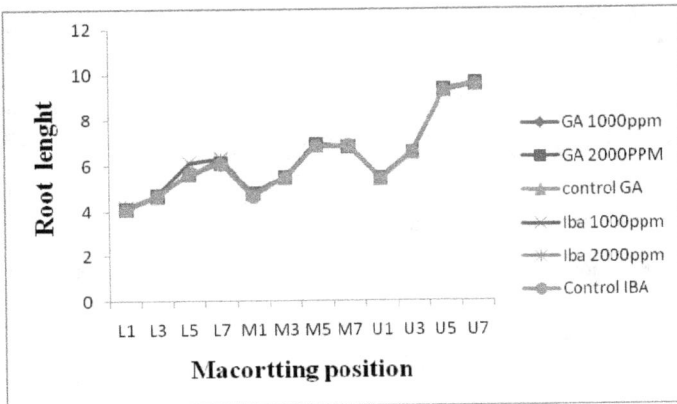

Figure 3 : the response of _D. edulis_ root length macortts to hormone types (GA and IBA), concentration and marcotting position

From (Figure 3) above its shows the relationship between hormonal type, hormone concentration and marcotting position, its shows that there is no visible different between the hormonal type and hormone concentration

The result of statistical analysis from table above shows that there were no significant differences (p>0.05) in the hormone types (GA and IBA) and at differences concetntrations., but there is significant differences (p>0.05) in the marcortting position (upper, middle and lower). (Appendix 2)

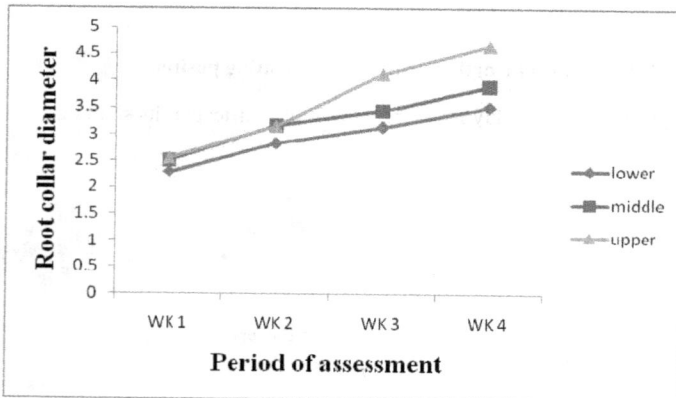

Figue 4 : the response of *D. edulis* root colar diameter macortts to marcotting position

From (Figure 4) the graph shows the weekly mean increament of root length and its the marcotting position (lower, middle and upper).

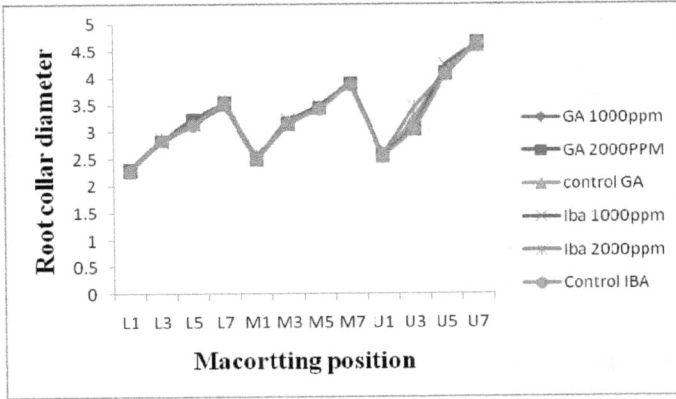

Figue 5 : the response of *D. edulis* root colar diameter macortts to hormone types (GA and IBA), concentration and marcotting position

Figure 5 above shows the relationship between hormonal type, hormone concentration and marcotting position, its shows that there is no visible difference between the hormonal type, hormone concentration on root collar diameter.

The result of statistical analysis from table above shows that there were no significant differences (p>0.05) in the hormone types (GA and IBA) and at differents concetntrations., but there were significant differences (p>0.05) in the marcortting position (Appendix 3)

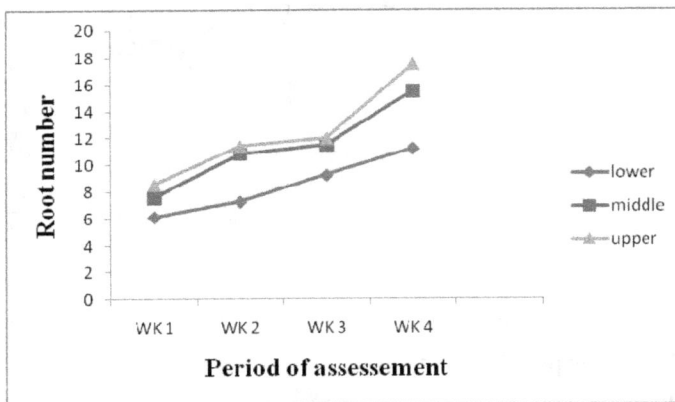

Figue 6 : the response of *D. edulis* macortts root number to marcotting position

From (figure 6) the graph shows the weekly mean increament of root length and the marcotting position (lower, middle and upper).

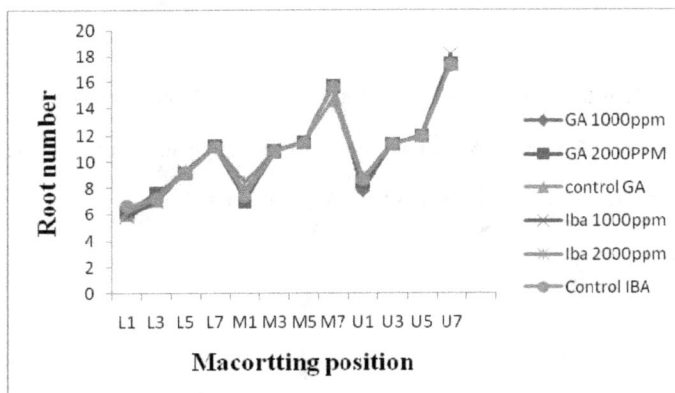

Figue 7 : the response of *D. edulis* root number macortts to hormone types (GA and IBA), concentration and marcotting position

Figure 7 above its shows the relationship between hormonal type, hormone concentration and marcotting position, its shows that there is no visible difference between the hormonal type, hormone concentration on root number

The result of statistical analysis from table above shows that there were no significant differences (p>0.05) in the hormone types (GA and IBA) and at differences concetntrations., but there was significant difference (p>0.05) in the marcortting position (Appendix 4).

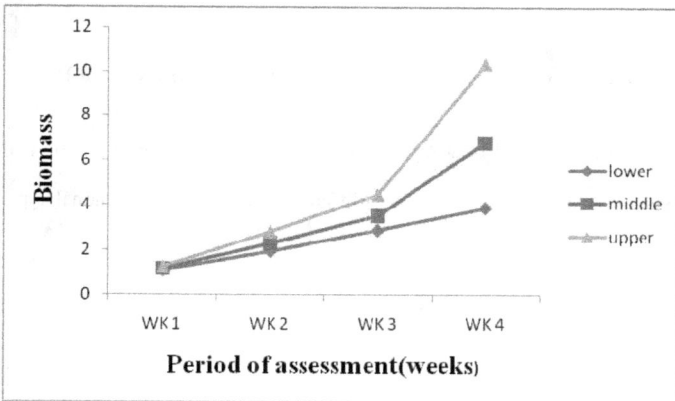

Figue 8 : The response of _D. edulis_ root biomass macortts to marcotting position

Figure 8 the graph shows the weekly mean increament of root length, its shows a clear distinction between the marcotting position (lower, middle and upper).

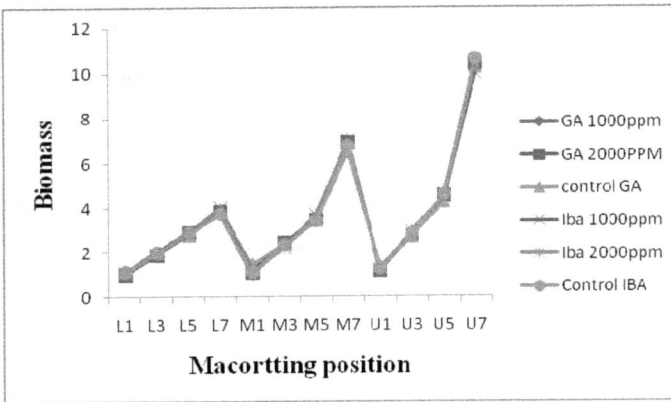

Figue 9 : the response of *D. edulis* root biomass macortts to hormone types (GA and IBA), concentration and marcotting position

Figure 9 above its shows the relationship between hormonal type, hormone concentration and marcotting position, its shows that there is no visible difference between the hormonal type, hormone concentration

Plate 1 : **Marcotted** *D. edulis*

Discussion

Phytosociology study of *D. edulis*

The measure of the coexistence and distribution of abundance of species, diversity has long been used to characterize the taxonomic structure of communities, this study's findings support assertion reported by GARRITY, (2006) that a woody tree species is justified to be used for multiple tree agroforestry if it supports that existence of other woody species around its niche. Hence, the observed associates of *D. edulis* in this study indicate great potential for agroforestry uses. In addition, the diversity of the *D. edulis* associates encountered in this study appeared to be appropriate, and the result of the study cobororate with in the findings of White (SABITTI EN 1992), who found that tree of similar evolutionary history tend to be adapted to a peculiar location and niche.

Vegetative propagation of *D. edulis*

The results from this study showed that there is no significant difference between the two hormone type and hormonal concentrations. The results corroborate with earliest reported by Isese, O. M. (2012), that hormonal treatment has no significant effects on the cuttings of *D. edulis*.

Results from this study shown that there is significant difference between the marcotting position for all the variables assessed. The upper section has the highest mean values than the lower section of the crown. The finding similars to the result of Ansari *et al* (1995) who reported that in *Dalbergia sissoo*, the cuttings from the apical end of the branch rooted better than the basal end, apical cuttings recorded higher mean percentage of rooted cuttings and root length than the basal cuttings. Upper section of the crown seems to have higher potential for vegetative propagation from this study having recorded mean values as found in *T. heterophylla* (Foster *et al*. 1984 and Larixdeciduas X L. kaempferi (Radosta *et al* 1994). The ability of cuttings to form roots is determined by the position of where the cutting is obtained, the juvenility of stock plant can also be an overriding. The results of this study indicated that positions had effects on roots development in *D. edulis*.with respects to roots length, root collar diameter, roots number and roots biomass. The result of this study however is in contrast to the statement of Hartman and Kester, (1983) that the best rooting is usually found from the basal portion of shoot. The differences in rooting responses with respect to marcotting position are greatly affected by he extent of lignifications, secondary thickening and chemical composition of plant tissues. Basal marcotted position could be to mature and highly lignified to develop roots than the apical position (Hartman *et al*. 1990). In woody plants, these differences in rooting due to marcotting position can be related to differences in chemical composition of the shoot, the age of the stem, carbohydrate accumulation or due to bud growth.

Conclusion

D.edulis is a multipurpose tree that provides a dual function of strengthening food security and Carbon sequestration in rural environments. This research is therefore pertinent to efforts towards indigenous fruit/nut tree domestication and incorporation to an agroforestry system. However, the research aimed at obtaining insights into the agroforestry potential of *D.edulis* and the possibilities of vegetative propagation of the species. This study showed that *D.edulis* has the potential of being used for multi-species agroforestry system and based on the large number highly economic values of the associated species, they have the potential of improving agroforestry practice in the area.

The results shown that marcotting is very possible for *D.edulis* and the marcort should be set at upper section of the main stem because its produces the best response to rooting in all the variables assessed, while the application of hormones may not be necessary for marcotting in *D. edulis*

The sustainable use of *D.edulis* can contributes to regional and local income generation, strengthens food security, improves health care and sequestrate carbon to fight climate change. Consequently, it is necessary to initiate studies aimed at getting insights on multipurpose tree species such as *D.edulis* on their bellow and above ground growth attributes and their potential ecosystem services to humanity to sustenance of both rural and urban livelihood, thus, An Agroforestry Innovation with multipurpose species should be advocated and encourage.

Reference

Ajibesin KK, Ekpo BJ, Bala DN, Essien EE, and Adesanya SA (2008). Ethnobotanical survey of Akwa Ibom state of Nigeria.*J. Ethnopharmacol.*, 115: 387-408.

Aminah, H., Nor Hasnita, R.M.N. & Hamzah, M. 2006. Effects of indole butyric acid concentrations and media on rooting of leafy stem cuttings of Shorea parvifolia and Shorea macroptera. Journal of Tropical Forest Science 8(1): 1-7.

Ansari. S. A, Kumar. P, Mandal, A. K. (1995). Effect of position and age of cuttings and auxins on induction and growth of roots in Dafbergia. Sissoo Roxb. Indian Forester 121, 201-206

Asaah E, Tchoundjeu Z, and Atangana A (2003). Cultivation and Conservation status of *Irvingia wombolu* in humid lowland forests of Cameroon. *Food, Agriculture & Environment*, 1: 251-256.

Atangana A.R., Tchoundjeu Z., Asaah E.K., Simons A.J. and Khasa D.P. 2006. Domestication of*Allanblackia fioribunda:* amenability to vegetative propagation. *Forest Ecology and Management* 237: 246--251.

Ayuk ET, Duguma B, Franzel S, KengueJ, Mollet M, Tiki-Manga T, Zekeng P (1999). *Use,* Management and economic potential of *Irvingia gabonensis*in the humid lowlands of Cameroon.*Forest Ecology and Management,* 113 1-9.

Barany M., Hammett A.L., Stadler K.M. and Kengni E. (2004). Non-timber forest products in the food security and nutrition of smallholders afflicted by HIV/AIDS in sub-Saharan Africa.*Forests, Trees and Livelihoods,* 14: 3–18.

De Klerk G.J (2000). Rooting of microcuttings: Theory and practice.*In Vitro Cell. Dev. Biol-Plant,* 38:415-422.

FAO (2005). The State of Food Insecurity in the World: Eradicating World Hunger. Key to Achieving the Millennium Development Goals. FAO, Rome pp 332

FAO (2007). Forest Resources Assessment, State of the world forest. Pp223

Foster, G. S., Campbell, R.K., and Adams, W.T. (1984). Heritability gain C effects in rooting of Western hemlock cuttings. Can J. Forest Resources. 14: 628-638.

GARRITY D 2006 Science-based agroforestry and the achievement of the Millenium Development Goals. *In*: Garrity D, Okono A, Grayson M, Parrott S (*eds*) World Agroforestry into the Future. World Agroforestry Centre, Nairobi, Kenya, pp 3-8

Gepts P, (2002). Review and Interpretation: A comparison between Crop Domestication, Classical Plant Breeding, and Genetic Engineering. Crop Science 42:1780-1780

Hartmann H.T and Kester DE (1983). Plant Propagation: Principles and Practices. Prentice- hall International, London pp33

Hartmann HT, Kester DE, Davies Jr FT, Geneve RL (2002). Plant propagation: Principles and practices. 7th Edition Upper Saddle River, New Jersey, USA PP65

Hartmann, H.T., Kester, D.E., Davies, F.T., Geneve, R.L., (2014). Hartmann and Kester's Plant Propagation: Principles and Practices, 8th ed. Prentice Hall, Englewood Cliffs, NJ.

Hartmann, H.T., Kester, E. and Davies, F.T. (1990). Plant Propagation. Principles and Practices.Fifth edition. Prentice-Halt International Edition? New Jersey. 645 pp.

Hendromono. (2007). Teknik pembibitan eboni dari anakan hasil permudaan alam. Jurnal Penelitian Hutan Tanaman, 4(2), 069 – 118

Howland P (1975). Vegetative propagation methods for *Triplochiton scleroxylon*K.Schum.*Proc. Symp.on Variation and Breeding Systems of*Triplochitons cleroxylon *K.Schum.*Federal Department of Forest Research, Ibadan, Nigeria.pp 99-109.

Igoli JO, Ogaji OG, Tor-Anyiin TA and Igoli NP (2005). Traditional medical practices among the Igede people of Nigeria. Part II. *Afr. J. Trad. CAM.*, 2: 134-152.

Isese, O. M. (2012) (influence of cutting position and clonal variation on rooting of *D.edulis) Unpublished* M.sc project, Forest Resources Management, University of Ibadan pp 34-41

Jaenicke, H. and Beniest, J. (2002) Vegetative Tree Propagation in Agroforestry. ICRAF, Nairobi, 1-30, 75-82.

Jiofack, T., C. Fokunang, N. Guedje, V. Kumeuze and E. Fongnzossie *et al.*, (2010). Ethnobotanical uses of medicinal plants of two ethnoecological regions of Cameroon. *Int. J. Med. Med. Sci.*, 2: 60-79.

Kengue J (2002). Safou. *Dacryodes edulis*, International Centre for Underutilise Crops, Southampton, UK.RPM Reprographics, Chinchester, England. 147p.

Kengue J and NyaNgatchou J (1990).Problème de la conservation du pouvoirgerminative chez les graines de safoutier (*Dacryodesedulis*).*Fruits*, 45 (4): 409 - 412

Kengue J and Tchio F (1994).Essais de bouturageet de marcottage du Safoutier (*Dacryodes edulis*). In: Kengue, J and NyaNgatchou J (eds) Le safoutier (*Dacryodes edulis*). Proceeding of Regional Seminar on the Importance of Safoutier, 4-6 October 1994, Douala, Cameroon.pp 80-98

Leakey RRB (2004). Physiology of vegetative reproduction. In J Burley, J Evans, and JA Youngquist (eds.) *Encyclopaedia of Forest Sciences*, Academic Press, London, UK. pp 1655-1668.

Leakey RRB (2005). Domestication potential of Marula (*Sclerocarya birrea*subsp*caffra*) in South Africa and Namibia: 3. Multi-trait selection. *Agroforestry Systems*, **64**:51-59.

Leakey RRB and Akinnifesi, F. K. (2008) Towards a domestication strategy for indigenous fruit tree in the tropics. CAB International, Wallingford, U K. P 28-49.

Leakey RRB, Mesén JF, Tchoundjeu Z, Longman KA, Dick JMcP, Newton A, Matin A, Grace J, Munro RC, Muthoka PN (1990) Low-technology techniques for the vegetative propagation of tropical trees. *Common Forest Rev*, 69:247-257.

Leakey RRB, Tchoundjeu Z, Schreckenberg K, Shackleton SC, Shackleton CM (2005). Agroforestry Tree Products (AFTPs): targeting poverty reduction and enhanced livelihoods. *Int J Agric Sustainability*, 3:1-23

Leakey, R.R.B., (1987). Clonal forestry in the tropics-- a review of developments, strategies and opportunities. Common. For. Rev., 66:61-75.

Mampouya P, Galamo G, Mialoundama F (1994). Recherche de nouveaux substratset influence des régulateurs de croissancesur le marcottageaérien du safoutier (*Dacryodes edulis* (G. Don) H. J. Lam). In: Kengue J. and NyaNgatchou J. (eds). Le safoutier, the African pear.Acte du séminaireRégionalsur la Valorisation du Safoutier, 4-6 October 1994. Douala, Cameroun. pp 72- 79.

Mapongmetsem PM (1994). Phénologieet mode de propagation de quelques essences locales à potentielagroforestier en zone forestière.Thèse de Doctorat 3ème Cycle, Université de Yaoundé I, 171 p

Mialoundama F, Avana M-L, Youmbi E, Mampouya PC, Tchoundjeu Z, Mbeuyo M, Galamo GR, Bell JM, Kopguep F, Tsobeng AC, and Abega J (2002). Vegetative propagation of *Dacryodes edulis*(G.Don) H.J. Lam by marcots, cuttings and micropropagation.*Forest, Trees and Livelihoods*, 12:85-96

Mng'omba SA, Akinnifesi FK, Sileshi G, Ajayi OC, Chakeredza C, Weston F, Mwase WF (2008). A decision support tool for propagating Miombo indigenous fruit trees of southern Africa.*Afr J Biotechnol,* 7:4677-4686

Naughton- Treves.L, Alix G. S. Chapman C. A (2011). Lessons about parks and poverty from a decade of forest loss and economic growth. Academy of sciences of the United State of America. 108: 139-190

Okwu DE and Nnamdi FU (2008). Evaluation of the chemical composition of *Dacryodes eduli s*and*Raphiahookeri*mann and wendl exudates used in herbal medicine in south eastern Nigeria.*Afr. J. Trad. CAM.,* 5: 194-200

Oni PI, Hall JB 2010 The phytosociology and ecological niche of P. biglobosa (Jacq) Benth: Implications for conservation and management. Nigerian Journal of Science 44: 17-28

Purugganan MD, Fuller DQ. Archaeological data reveal slow rates of evolution during plant domestication, *Evolution,* 2011, vol. 65 (pg. 171-183)

Radosta, P., Paques, L.E and Verger, M. (1994). Estimation of genetic and non-genetic parameters for rooting traits in hybrid larch. Silvae Genet. 43: 108-114

Sabitti EN, Cobbina J 1992 Parkia biglobosa: a potential multipurpose fodder in tree legume in West Africa. International Tree Crop Journal 7 (3): 113-139. DOI: http://dx.doi.org/10.1080/0143569 8.1992.9752911

Schreckenberg K, Awono A, Degrande A, Mbosso C, Ndoye O and Tchoundjeu Z (2006). Domesticating indigenous fruit trees as a contribution to poverty reduction.*Forests, Trees and Livelihoods*, 16:35–51

Tchoundjeu Z, Atangana A, Asaah E, Tsobeng A, Facheux C, Foundjem D, Mbosso, C, Degrande A, Sado T, Kanmegne J, Mbile P, Tabuna H, Anegbeh P, Useni M (2008). Domestication, utilization and marketing of indigenous fruit trees in West and Central Africa. In: Akinnifesi FK, Leakey RRB, Ajayi OC, Sileshi G, Tchoundjeu Z, Matakala P, Kwesiga FR (2008). Indigenous fruit trees in the tropics: domestication, utilization and commercialization. pp. 171-185. CAB International. Wallingford.

Tchoundjeu, Z, Degrande A, Leakey RRB, Simon, AJ, Nimino G, Kemajou E, Asaah E, Facheux C, Mbile P, Mbosso C, Sado T and Tsobeng A (2010). Impact of participatory tree domestication on farmer livelihoods in west and central Africa.*Forests, Trees and Livelihoods,* **19**: 219-234.

Tchoundjeu, Z. and Leakey, R.R.B. (2006). Vegetative propagation of African mahogany: Effects of auxin, node position, leaf area and cutting length. *New Forests* 21:125-136.

Teklehaimanot, A., McCord, G. C., & Sachs, J. D. (2007). Scaling up malaria control in Africa: An economic and epidemiological assessment. American Journal of Tropic Medicine and Hygiene, 77 , 138–144.

Verheiji EWM (1991). Introduction. In: Verheij E.W.N. and Coronel, R.E.9eds) Edible Fruits and Nuts. Plan Resources of South East Asia No. 2 Pudoc, Wageningen, The Netherlands. pp. 15-56.

Waruhiu AN, Kengue J, Atangana AR, Tchoundjeu Z and Leakey RRB (2004). Domestication of *Dacryodes edulis*.2. Phenotypic variation of fruits in 200 trees from four populations in the humid lowlands of Cameroon, in *Food, Agriculture and Environment,* 2:340-346

Weber J. C,.Sotelo. C.M., Vidaurre H., Dawson I. K,. and Simons A.J (2004). Participatory Domenstication of agroforestation trees: An example from the Peruvian Amazon peruviana. Development practice. 11: 425- 433.

Wiersum KF, (2008). Domestication of trees or forest: development pathways for fruit tree production in south East Asia. p. 708-3 in FK Akinnifesi, RRB Leakey, OC Ajayi, G Sileshi, Z Tchoundjeu, P Matakala and F Kwesiga (eds.) *Indigenous Fruit Trees in the*

Tropics: Domestication, Utilization and Commercialization, CAB International, Wallingford, UK

Zomer RJ, Trabucco A, Coe R and Place F (2009). Trees on farm: Analysis of global extent and geographic patterns of agroforestry. *ICRAF Working Paper No 89*, World Agroforestry Centre, Nairobi, Kenya, 63pp

Zwane AP (2007). Does poverty constrain deforestation? Econometric evidence from Peru.Journal of DevelopmentEconomics, 84: 330-349.

APPENDIX

Appendix 1:   ANOVA Table for Root length

Source	Df	Sum of Squares	Mean Square	F	Sig.
Hormone types (HT)	1	0.00	0.00	0.00	0.95^{ns}
Hormone concentration (HC)	2	2.38	1.19	1.52	0.23^{ns}
Marcotting postion (MP)	2	106.15	53.07	67.70	0.00^*
HT*HC	1	1.03	1.03	1.31	0.26^{ns}
HT*MP	2	0.03	0.02	0.02	0.98^{ns}
HC*MP	4	2.04	0.51	0.65	0.63^{ns}
HT*HC*MP	2	0.98	0.49	0.62	0.54^{ns}
Error	75	58.79	0.78		
Total	89	171.40			

*=significant ($p<0.05$)
Ns = not significant different

Appendix 2:   ANOVA Table for Root collar diameter

Source	Sum of Squares	df	Mean Square	F	Sig.
Hormone types (HT)	0.001	1	0.001	0.003	0.955^{ns}
Hormone concentration (HC)	0.392	2	0.196	1.148	0.323^{ns}
Marcotting postion (MP)	6.646	2	3.323	19.455	0.00^*
HT*HC	0.027	1	0.027	0.156	0.694^{ns}
HT * MP	0.235	2	0.117	0.687	0.506^{ns}
HC * MP	0.135	4	0.034	0.198	0.939^{ns}
HT * HC * MP	0.208	2	0.104	0.609	0.547^{ns}
Error	12.81	75	0.171		

Total	973.241	90			

*=significant (p<0.05), Ns = not significant different

Appendix 3:   ANOVA Table for Root number

Source	Sum of Squares	df	Mean Square	F	Sig.
Hormone types (HT)	2.628	1	2.628	2.014	0.16^{ns}
Hormone concentration (HC)	0.077	2	0.039	0.03	0.971^{ns}
Marcotting postion (MP)	254.117	2	127.058	97.381	0.00^*
HT * HC	0.234	1	0.234	0.18	0.673^{ns}
HT * MP	0.34	2	0.17	0.13	0.878^{ns}
HC * MP	0.842	4	0.21	0.161	0.957^{ns}
HT * HC * MP	0.156	2	0.078	0.06	0.942^{ns}
Error	97.856	75	1.305		
Total	10596.25	90			

*=significant (p<0.05)
Ns = not significant different

Appendix 4:   **ANOVA Table for Biomass assessment**

Source	Sum of Squares	df	Mean Square	F	Sig.
Hormone types (HT)	0	1	0	0.002	0.968^{ns}
Hormone concentration (HC)	0.204	2	0.102	0.441	0.645^{ns}
Marcotting postion (MP)	78.258	2	39.129	169.149	0.00^*
HT * HC	0.059	1	0.059	0.254	0.616^{ns}
HT * MP	0.005	2	0.002	0.01	0.99^{ns}

HC * MP	0.549	4	0.137	0.594	0.668^{ns}
HT * HC * MP	0.049	2	0.024	0.105	0.9^{ns}
Error	17.35	75	0.231		
Total	1255.703	90			

*=significant (p<0.05)
Ns = not significant different

www.ingramcontent.com/pod-product-compliance
Lightning Source LLC
Chambersburg PA
CBHW061301220326
41599CB00028B/5735